AF575346

FRUITS WE EAT

Katherine Rawson

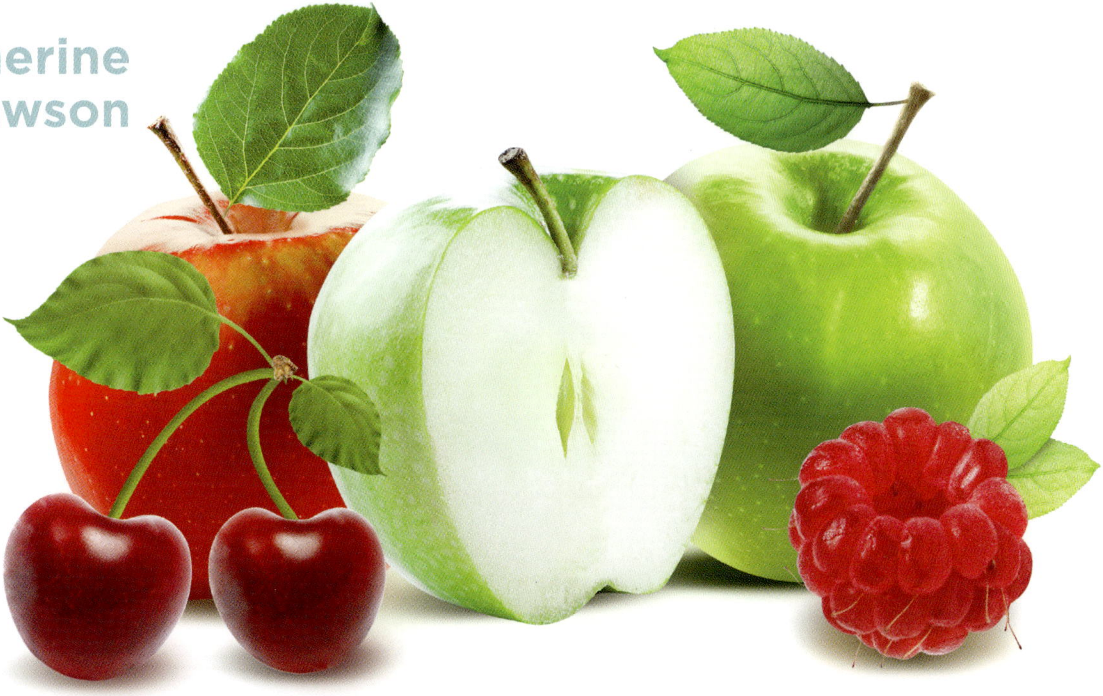

EZ READERS

Creating Young Nonfiction Readers

EZ Readers lets children delve into nonfiction at beginning reading levels. Young readers are introduced to new concepts, facts, ideas, and vocabulary.

Tips for Reading Nonfiction with Beginning Readers

Talk about Nonfiction
Begin by explaining that nonfiction books give us information that is true. The book will be organized around a specific topic or idea, and we may learn new facts through reading.

Look at the Parts
Most nonfiction books have helpful features. Our *EZ Readers* include a Contents page, an index, and color photographs. Share the purpose of these features with your reader.

Contents
Located at the front of a book, the Contents displays a list of the big ideas within the book and where to find them.

Index
An index is an alphabetical list of topics and the page numbers where they are found.

Photos/Charts
A lot of information can be found by "reading" the charts and photos found within nonfiction text. Help your reader learn more about the different ways information can be displayed.

With a little help and guidance about reading nonfiction, you can feel good about introducing a young reader to the world of *EZ Readers* nonfiction books.

Mitchell Lane
PUBLISHERS

2001 SW 31st Avenue
Hallandale, FL 33009
www.mitchelllane.com

First Edition, 2021.

Author: Katherine Rawson
Designer: Ed Morgan
Editor: Morgan Brody

Names/credits:
Title: Fruits We Eat / by Katherine Rawson
Description: Hallandale, FL :
Mitchell Lane Publishers, [2021]

Series: Plant Parts We Eat
Library bound ISBN: 978-1-58415-046-6
eBook ISBN: 978-1-58415-049-7

EZ Readers is an imprint of Mitchell Lane Publishers.

Photo credits: Freepik.com, Shutterstock

Contents

Plants have fruit.
The fruit holds the seeds.
The fruit **protects** the seeds until they are **ripe**.

We eat many kinds of fruit.
Apples are sweet fruit.
Apples grow on trees.
Each apple has several seeds inside.

Did You Know?

There are over 7,500 different kinds of apples grown around the world.

Cherries are sweet fruit.
Cherries grow on trees, too.
Each cherry has one seed inside.
It is called a **pit**.

Did You Know?

Some cherries are sweet and some are sour. Sour cherries make good pies.

Berries are sweet fruit. Many berries grow on bushes. Strawberries grow on the ground. Berries have many tiny seeds.

Did You Know?

Raspberry bushes are covered with thorns. The thorns keep animals from eating the plant.

Sweet fruit makes a good snack.
Tomatoes are not sweet.
But tomatoes are fruit.

Did You Know?

Tomatoes come in many colors and sizes.

Each tomato holds many small seeds. Tomatoes taste good in salads, sauces, and soups.

Cucumbers are fruit, too. They have seeds inside. They taste good in salads and sandwiches.

Tomatoes and cucumbers grow on **vines.**

All kinds of fruit are good to eat.

Glossary

pit
A single seed

protect
Keep safe

ripe
Finished growing

vine
A climbing stem that cannot support itself

Sources

https://www.cropsreview.com/functions-of-fruits.html

https://www.britannica.com/science/fruit-plant-reproductive-body

https://researchfrontiers.uark.edu/why-do-plants-have-thorns/

https://learningbynature.org/lessons-from-thorn/

https://extension.illinois.edu/apples/facts.cfm (# of varieties)

https://homeguides.sfgate.com/prepare-cherry-pits-planting-germination-68796.html

https://www.thetreecenter.com/berries-grow-trees/

Further Reading

Web Pages:
Read more about fruit
https://extension.illinois.edu/gpe/case1/c1facts2e.html

Read more about plant parts
http://www.mbgnet.net/bioplants/parts.html

See photos of different kinds of fruit
https://www.halfyourplate.ca/fruits-and-veggies/fruits-a-z/

Books:
A Tree Is a Plant
By Clyde Robert Bulla
(HarperCollins, 2016)

The Amazing Life Cycle of Plants
By Kay Barnham
(B.E.S. Publishing, 2018)

Index

About the Author

Katherine Rawson loves growing, cooking, and eating vegetables. She also loves writing. So, she thought it would be a great idea to write books about plants we eat. She likes eating all kinds of fruit. Her favorite summer activity is picking wild berries.